YOUR KNOWLEDGE HAS VALUE

- We will publish your bachelor's and
  master's thesis, essays and papers

- Your own eBook and book -
  sold worldwide in all relevant shops

- Earn money with each sale

Upload your text at www.GRIN.com
and publish for free

Bibliographic information published by the German National Library:

The German National Library lists this publication in the National Bibliography; detailed bibliographic data are available on the Internet at http://dnb.dnb.de .

This book is copyright material and must not be copied, reproduced, transferred, distributed, leased, licensed or publicly performed or used in any way except as specifically permitted in writing by the publishers, as allowed under the terms and conditions under which it was purchased or as strictly permitted by applicable copyright law. Any unauthorized distribution or use of this text may be a direct infringement of the author s and publisher s rights and those responsible may be liable in law accordingly.

Imprint:

Copyright © 2012 GRIN Verlag, Open Publishing GmbH
Print and binding: Books on Demand GmbH, Norderstedt Germany
ISBN: 978-3-668-03110-4

This book at GRIN:

http://www.grin.com/en/e-book/304414/computer-assisted-robotic-surgery-and-telesurgery

Ty Lantz

Computer-assisted robotic surgery and telesurgery

GRIN Publishing

GRIN - Your knowledge has value

Since its foundation in 1998, GRIN has specialized in publishing academic texts by students, college teachers and other academics as e-book and printed book. The website www.grin.com is an ideal platform for presenting term papers, final papers, scientific essays, dissertations and specialist books.

Visit us on the internet:

http://www.grin.com/

http://www.facebook.com/grincom

http://www.twitter.com/grin_com

Tyson Lantz

University of Nebraska
at Omaha

Computer-Assisted
Robotic Surgery and
Telesurgery

12/4/2012

Computer-Assisted Robotic Surgery
and Telesurgery
Tyson H. Lantz
University of Nebraska at Omaha
12/5/2012

Table of Contents

Figures and Tables

Executive Summary

The subject of this report is an introduction to computer-assisted robotic surgery and telesurgery.

The scope of this report includes information on the following major areas:

- The foundation of computer-assisted robotic surgery
- The components of a computer-assisted robotic surgical system
- The advantages and disadvantages of computer-assisted surgery
- The explanation of telemedicine and telesurgery
- The technical issues during telesurgery
- The technical and legal issues with telesurgery
- The first telesurgery procedure
- Two case studies involving computer-assisted surgery
- Telesurgery as an educational tool
- Recent advances and future applications of computer-assisted robotics in health care
- Future applications of computer-assisted robotics in telesurgery

The purpose of this report is to explore and describe the innovation of computer-assisted robotic surgery and telesurgery in medicine.

The foundation of computer-assisted robotic surgery is derived from laparoscopic surgery, a technique that uses multiple tiny incisions in the place of one large incision. Components of a surgical system include the surgeon's console, the patient-side cart, and the image-processing and insufflations stack. The robot is operated by the surgeon's console which has hand and foot controls that are used to operate the patient-side cart. The patient-side cart contains four robotic arms, all of which can be used to operate a surgical instrument, hold a camera used to view the operative site, or hold a retractor. The image-processing and insufflations stack contains a camera control unit, an image-recording device, an insufflator used to blow air or gas into a body cavity, and a monitor used by operating assistants.

Computer-assisted robotic surgery offers numerous advantages over other surgical approaches. The human wrist only offers four degrees of movement, but the EndoWrist of the *daVinci* Surgical System offers its users seven degrees of movement. Furthermore, the *daVinci* Surgical System offers its users three-dimensional image viewing as compared to traditional laparoscopic cameras which offer only a two-dimensional image. While computer-assisted robotic surgery offers advantages, it doesn't come without a few disadvantages. Institutionally, costs include the purchase of a surgical system; the *daVinci* surgical system is one million dollars. Additionally, maintenance costs run one-hundred thousand dollars a year, and training for staff costs another quarter of a million dollars a year. Other disadvantages include the physician learning curve as well as the initial lengthy set-up times for operating room staff.

Telemedicine is a product of the 21st century telecommunication and information technologies, and because it eliminates distance barriers, telemedicine is capable of providing clinical health care access to patients in distant rural communities. By utilizing a combination of

Executive Summary

telecommunication and information technologies along with surgical robotics, a doctor has the ability to perform surgery on a patient even though the patient is not physically in the same location during a process known as telesurgery, a subset of telemedicine. Issues involved with telesurgery include malpractice liability and connectivity issues that affect operation of the robot during a procedure.

The first telesurgery performed was in 1991, on a sixty-eight year-old female patient in Strasbourg, France, while the surgeon was over 4,000km away in New York City, NY. The procedure was completed without incident and the patient made a full recovery. Through new teaching practices such as linking two surgeon's consoles together for the purposes of educating, it is believed that "telesurgery might eventually improve the standard of surgical care throughout the world" (Marescaux et al., 2002).

Advances in technology in the 20th and 21st centuries have grown at an extraordinary rate, and these technological advances have both revolutionized and forever changed the manner in which people conduct their daily lives. Among these advances, those in the area of medicine have probably had the most profound and far-reaching effect for people in terms of improving their quality of life. Throughout their lives, most people have something happen to themselves, a family member, or a friend which requires medical attention and ultimately the need for surgical intervention. In the fast-paced society in which people live today, people want to return to living their daily lives as quickly as possible, and the use of computer-assisted robotic surgery offers them the least invasive surgical approach with the best possible outcome should surgery be required. As technology and telecommunications progress, people will see the use of telemedicine and telesurgery become even more entrenched in health care practice. Who knows what phenomenal breakthroughs and medical advancements are just beyond the horizon? Only time alone will reveal what remarkable innovations mankind will see in the future which include computer-assisted robotics and telesurgery in the practice of medicine.

Computer-Assisted Robotic Surgery and Telesurgery
Tyson H. Lantz
University of Nebraska at Omaha

The Latin phrase *Primum non nocere,* which translates to "First, do no harm", is a common ethical lesson taught to medical students, and it is a fundamental principle to be followed in practicing medicine. The importance of this statement becomes even more emphatic when one considers that the practice of medicine today is becoming continually more reliant on technology to aid in the diagnosis and treatment of diseases and conditions. Technology is an everyday aspect of people's daily lives, and their lives continue to be improved with computers and technological advancement. For example, the use of technology, specifically computers, has been in use in the medical setting for fifty years, however, within the last twenty years, computers have started to make their way into the operating room at an ever increasing rate. As a result of this increased use in the operating room, computers are becoming highly valuable and crucial members of the surgical team.

By combining clinical decision support systems with patient-specific data, physicians now have the ability to use intelligent devices and technologies to create a perioperative zone of safety for patients. Intuitive Surgical (2012), a computer-assisted robotic surgical system developer and manufacturer, explains that the *daVinci* Surgical System has the capabilities necessary to "perform delicate and complex operations through a few tiny incisions with increased vision, precision, and dexterity and control". This translates into decreased blood loss, reduced tissue trauma, reduction in length of stay at the hospital, and increased postoperative recovery for the patient. Furthermore, in situations where the surgeon and patient are physically in different locations, some surgical systems can be used by the surgeon to perform surgery on the patient, using an innovative approach known as telesurgery. Computer-assisted robotic surgery is now becoming an accepted standard of practice in the operative setting. Further refinement of its use and continued developments in the area of telesurgery are the future of today's medicine. The following presents an in depth investigation and discussion of the evolution of computer-assisted robotic surgery and telesurgery in medicine.

The Foundation of Computer-assisted Robotic Surgery

Modern surgery has come a long way from bloodletting, the practice of draining a person's blood to cure an ailment, dating as far back as the fifth century BC (Sharp, 2001). Previous to surgery with robots, years of education and development of surgical practice and technique was needed. As a result of this development of surgical practice and technique, tremendous benefits have been yielded for those people requiring surgery. In short, the realm of surgery has evolved from that which is known as invasive surgery to that which is known as laparscopic surgery, also referred to as minimally invasive surgery. Computer-assisted robotic surgery takes minimally invasive surgery and builds upon the concept.

Invasive and Minimally Invasive Surgery

Before laparoscopic and computer-assisted robotic surgery, invasive surgery involved a large incision, also referred to as an "open" procedure or laparotomy, where the entire cavity of the patient was exposed for the surgeon. In comparison, the premise behind laparoscopic surgery or minimally invasive surgery is to access the body cavity through multiple tiny incisions, usually no bigger than a half-inch in diameter. Ports, also known as trocars, are placed into these small incisions, producing channels to allow access to the inside of the patient. These channels provide a way for longer instruments to be introduced and used in the body. Additionally, a small camera, known as a laparoscopic camera, is placed through one of the ports. It is used to transmit an image onto a television monitor for the surgeon and assistants to view. In this manner, the laparoscopic camera becomes the surgeon's eyes because the surgeon is not able to see directly into the patient without the traditional large incision.

Components of a Computer-Assisted Robotic Surgical System

As stated previously, computer-assisted robotic surgery takes laparoscopic or minimally invasive surgery and builds upon the concept. During computer-assisted robotic surgery, the surgeon is able to expand the capabilities within the operative field while doing so in an even less invasive way than traditional laparoscopic surgery. Satava explains the concept of a surgical system by stating that it "is not a machine, it is an information system with arms" (2005). In essence, "by using the robot, the surgeon looks at the video image (the electronic representation of the organs) and moves the handles which send electronic signals (information) to the tips of the instruments – surgery becomes a flow of information" (Satava, 2005). One such robotic surgical system, the *daVinci* (as represented in the drawing below), is composed of three components which are the surgeon's console, the patient-side cart, and the image-processing and insufflations stack. The following is a brief discussion of each of these components.

Surgeon's Console

Once again, a surgical system is not a machine, and it is not artificially intelligent, i.e., it does not

have a mind of its own. Therefore, a surgical system needs the control of a surgeon to perform any duties; hence, it is a master-slave unit. The surgeon's console (master unit) is placed within the surgical suite but away from the operating table. A three-dimensional image from the stereoscopic endoscope, which is an instrument inserted into an internal organ of the body cavity to allow viewing of its interior, is projected into the console, magnified at ten times the normal viewing capabilities of the human eye for precise viewing of the operative site (Murphy et al., 2008). The placement of the surgeon's thumb and forefinger at the telemanipulators (controls) in the surgeon's console and subsequent movements of the telemanipulators are then transferred to movements of the robotic 'arms'. The foot controls at the surgeon's console provide power to a diathermy, which is used to aid in the cutting and cauterization of tissue, as well as to control other energy sources used in surgery.

Patient-side Cart and Image-Processing and Insufflations Stack

The robotic 'arms' are located on the patient-side cart (slave unit) which is placed at the operating table. The patient-side cart has the capability of providing up to four robotic 'arms', all of which can hold a high-resolution three-dimensional stereoscopic endoscope, instrument or retractor, a device used to separate an incision or wound or to hold tissue in place. The final piece of equipment is the image-processing and insufflations stack which contains the camera-control unit for three-dimensional image processing, an image-recording device, a laparoscopic insufflator used to blow air or gas into a body cavity, and a monitor with two-dimensional viewing for assistants.

Advantages of Computer-Assisted Robotic Surgery

Computer-assisted robotic surgery offers multiple benefits over traditional invasive and laparoscopic surgery. These advantages include those specific to the surgeon such as increased wrist dexterity, motion control, and vision. Computer-assisted robotic surgery also allows for the use of sturdier instrumentation better suited to laparoscopic surgery in morbidly obese patients. In addition, the *daVinci* surgical system has the ability to "import the patient's holomer and the surgeon can perform preoperative planning, surgical rehearsal or even surgical simulation for training and assessment" (Satava, 2005). As explained by VirtualSoldier.us, "The holomer (HOLO-graphic M-edical E-lectronic R-epresentation) is a three dimensional holographic digital image of a specific person" (2012). Physicians have the ability "to interact with the holomer as if it were the patients themselves" (VirutalSoldier.us, 2012).

Surgeon's Advantages

EndoWrist. Computer-assisted robotic surgery has numerous advantages over traditional laparoscopic surgery. During computer-assisted robotic surgery, the surgeon is able to expand the capabilities within

the operative field while doing so in a less invasive way than traditional laparoscopic surgery. Traditional laparoscopic instruments only offer surgeons four degrees of freedom, while the EndoWrist, which is available on the *daVinci* computer-assisted robotic surgical system, offers seven degrees of freedom (Murphy, Hall, Rong, Goel, Costello, 2008; Intuitive Surgical, 2012). EndoWrist instruments offer the surgeon natural dexterity, along with providing even greater range of motion that is naturally exhibited by human wrist articulations much like open surgery (Intuitive Surgical, 2012). Unlike traditional laparoscopic instruments, Intuitive Surgical explains that the internal cables of EndoWrist instruments offer maximum responsiveness, allowing rapid and precise suturing, dissection and tissue manipulation (2012). Traditional laparoscopic instruments have a 'fulcrum' effect, which occurs when the tip of the instrument's movements are opposite the direction of the surgeon's hand movements. The 'intuitive' nature of computer-assisted robotic surgery cancels this effect, allowing the instruments to move in the precise direction of the surgeon's hand movements in the surgeon's console (Murphy et al., 2008).

  Motion Scaling. Dr. Ayal M. Kaynan explains, "In standard laparoscopy, distance of the tissue structure from the port site on the abdomen causes amplification of motion at the instrument tip" (2009). The slightest motion outside of the patient can translate into a relatively large motion inside the operative field. Motions can be scaled up to a five-to-one ratio by filtering and de-amplifying the surgeon's hand movements at the console. In doing so, the surgeon must move the manipulators at the console five inches for the tip of the instrument to move one inch. As stated by Dr. Ayal M. Kaynan, "This eliminates natural hand tremor entirely, allows the surgeon to target tissues with much greater ease, and gives the surgeon a certain finesse that surpasses human capabilities in both the open and standard laparoscopic realms" (2009). In addition, the *daVinci* Surgical System uses a motion filter to isolate the frequency at which hand tremors occur, "eliminating unintended movements caused by human tremor" (Hirano, Ishikawa, Watanabe, 2010).

  Three-Dimensional Imaging. With conventional laparoscopes, the image viewed by the surgeon and assistants are limited to two-dimensional vision. The *daVinci* robotic laparoscope is actually comprised of two separate lenses that are used in conjunction to give the perspective of a three-dimensional image which mimics natural vision and gives true precision depth (Kaynan, 2009; Intuitive Surgical, 2012). The high-resolution three-dimensional stereoscopic vision allows the surgeon to target tissues properly by not having to make inferences about spatial relations, which typically happens with the traditional two-dimensional image provided by a standard laparoscope. Many times a 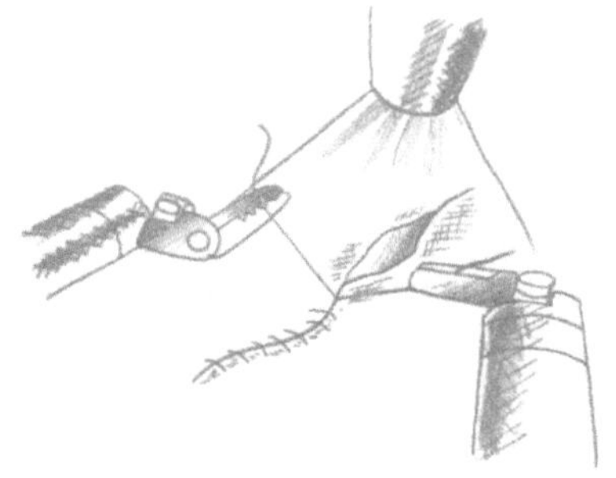 surgeon will make small test movements to either confirm or refute the inferences made about spatial relationships; the *daVinci* robot all but eliminates the need to make test movements.

Morbidly Obese Patients

In the morbidly obese, traditional laparoscopic instruments tend to bend. This problem is due to the increased pressure loads that must be exerted on the instruments by the surgeon. Precise operative technique becomes even more difficult with patients of a body mass index (BMI) greater than 60 kg/m^2 because of the added torque that must be exerted on the instruments (Jacobsen, Berger, Horgan, 2003). This dilemma is alleviated by the use of instrumentation which is more rigid and created for such situations. As explained by Jacobsen et al., "the stiffer instrumentation eliminates bending of the instruments, and the mechanical power provided by the robot eliminates surgeon fatigue and allows a coordinated, precise manipulation of tissues" (2003).

Disadvantages of Computer-Assisted Surgery

As previously described, computer-assisted robotic surgery offers a multitude of benefits over traditional invasive and laparoscopic surgery. However, it does not come without some disadvantages or drawbacks. These include those specific to the institution, patient, and surgeon as well as the operating room staff. Among these are cost, possible suture failure, physician learning curve, and setup time in preparing the robot for surgery. While none of these are potentially life-threatening, they all pose areas of concern which can be weighed as possible drawbacks of computer-assisted robotic surgery.

Institutional

Even though computer-assisted robotic surgery offers numerous benefits, there are some drawbacks that are related to the physical system as well as disadvantages encountered during operation. Hospitals can easily be deterred by the implementation, operating, maintenance and training costs associated with computer-assisted robotic surgery. While many health care facilities are not equipped financially to incur the large costs associated with the use of computer-assisted robotic surgery, there are over 1,840 *daVinci* Systems installed in over 1,450 hospitals worldwide (Intuitive Surgical, 2012). Data put together by Diana Gehardus show that initial costs for the *daVinci* Surgical System are in the neighborhood of one million dollars, with an annual maintenance cost of one-hundred thousand dollars and physician training running a quarter of a million (as cited in Computer Motion, Inc., 2002; Intuitive Surgical, Inc., 2002; Value and Feasibility, 2002). Gehardus also states each surgical instrument for the *daVinci* must be replaced after ten cases (procedures) at a price of two thousand dollars per instrument (as cited in Stark, 2002).

Patient

In addition to the costs associated with the *daVinci* System, another drawback is in the area of possible suture failure. When transitioning from open surgery, as well as laparoscopic surgery to computer-assisted robotic surgery, the biggest difference that a surgeon will incur is no tactile feedback to movement of the telemanipulators. As reported by Hirano et al., "Some studies have shown that the higher risk of breaking sutures when using the robot can be attributed to the absence of tactile feedback in the robotic system" (2010). If the surgeon does not break the suture, lack of tactile feedback could also lead to the stressing of the tensile strength of the suture, potentially leading to premature failure. Thus, surgeons are even more reliant on the three-dimensional picture provided by the high definition camera when suturing. It has also been noticed that if proper care is not taken to monitor the position and tension of retraction devices that are held in a fixed position by the robot, the potential for unintended tissue trauma can occur

(Newlin, Maikami, Melvin, 2004). It should be noted that although *daVinci* Surgical Systems at this time does not offer tactile feedback with their system, Surgeon's Operating Force-feedback Interface Eindhoven, or Sofie and NeuroArm provide tactile feedback on their computer-assisted robotic surgical systems (Coxworth, 2010; Hawaleshka, 2002).

Surgeon

As stated previously, costs incurred by a health care facility to train physicians to use the *da Vinci* System totaled a quarter of a million dollars, however, training also presents other challenges in addition to the financial challenge. As one might expect, there is a learning curve associated with adjusting to the surgeon's console, which includes the three-dimensional screens for viewing, telemanipulators and foot controls. Training centers are located throughout different locations within the United States, where training involves picking up objects with robotic arms by maneuvering the telemanipulators, as well as operating on human and animal cadavers among other exercises (Gehardus, 2003). Rigorous training of up to and over forty hours is required for the physician to even be considered familiar with the operation of the computer-assisted robot. Gehardus also states that "the surgeon may need to operate on 12 to 18 patients using the technology before he or she can feel comfortable and can perform the operation within standard times" (as cited in Meadows 2002).

Operating Room Staff

With the addition of new and unfamiliar technology into the operating room, sterile surgical team members as well as non-sterile surgical team members initially incur lengthy setup times associated with preparing the robot for a procedure. For example, at the University of Illinois, initial setup time on the *daVinci* robot for a Roux-en-Y gastric bypass procedure was thirty-five minutes (Jacobsen et al., 2003). However, Joacobsen et al., went on to say that after two hundred cases with the *daVinci* Surgical System, team members became so proficient at preparing the robot, seven minutes became the average time added onto each case (2003). Also, at the New York Medical College, surgeons reported similar times associated with the setup of the robot, with a six to eight minute window becoming the average setup time (Jacobsen et al., 2003). Nevertheless, the learning curve associated with the operating staff to become efficient in preparing the robot for a procedure is definitely a disadvantage of computer-assisted robotic surgery.

Telemedicine and Telesurgery

Telemedicine is yet another new aid in providing improved health care for patients. Telemedicine is a product of the 21st century telecommunication and information technologies, and because it eliminates distance barriers, telemedicine is capable of providing clinical health care access to patients in distant rural communities. The American Telemedicine Association formally defines telemedicine by stating it "is the use of medical information is[sic] exchanged from one site to another via electronic communications to improve a patient's clinical health status" (2012). "The use of telemedicine for different clinical problems include large distances between patients and specialists, isolated health professionals requiring special support and/or education, and situations where there is no other alternative, for example space flight or patients at sea" (Smith, Bensink, Stillman, Caffery, 2005). Telemedicine has grown to encompass "a growing variety of applications and services using two-way video, email, smart phone, wireless tools and other forms of telecommunications technology" (American Telemedicine Association, 2012). This two-way communications between patient and care provider can facilitate "and deliver a host of health-related information and health care service" (Lee, Harada, 2011).

Telemedicine can be broken down into a wide array of medical services, one of which is the area known as telesurgery. The first 'raw' definition of 'telesurgery' is the use of telecommunication technology to aid in the practice of surgery (Hanly, Marohn, Schenkman, Miller, Marchessault, & Broderick, 2005). Hanly et al., explains that this broad definition "encompasses everything from pre-, intra-, and postoperative teleconsultation and tele-evaluation of the patient, to intraoperative telementoring, teleproctoring and telemonitoring, to telemanipulation and telepresent robotic surgery" (as cited in Whitten, Mair, 2004; Lee, Broderick, Haynew, Bagwell, Doarn, Merrill, 2003; Rodas, Latifi, Cone, Broderick, Doarn, Merrill, 2002). By utilizing a combination of telecommunication and information technologies along with surgical robotics, a doctor has the ability to perform surgery on a patient although they are not physically in the same location. This innovative approach to surgery offers valuable medical care to patients who otherwise might not receive the necessary and even life-saving medical attention they require.

Technical Issues During Telesurgery

When discussing telesurgery, there is an area of potential technical issues that could arise during a procedure. Some of the technical issues presented are similar to those incurred while playing online games or streaming movies and other content through an active internet connection. Technical issues endured during remote telesurgery can be so minute that the surgeon does not notice any difference in picture or control. On the other hand, some technical issues can be severe enough to make surgery impossible.

Technical Issues Defined

The objective of telemedicine is to replicate on-site services at a distant location through the use of telecommunications and supporting technologies. Hanley et al., states that "the ability to capture, compress, enhance, and transmit digital video, provides the technology necessary to make telesurgery a reality" (2005). Telemedicine's judged successfulness is heavily weighted on how such communications and technologies are able to replicate on-site services correctly and without incident. With this in mind, there is need to define technical terms that affect telesurgery. These terms include *control latency, lag, round-trip delay, bandwidth,* and *jitter,* and the following is an explanation of each.

- **Control latency.** The delay between when a remote surgeon moves the telemanipulators and when the surgical instrument actually moves inside of the patient is defined as control latency. Hanley et al., further defines control latency by stating, "this time is the sum of the delays inherent to digitization of the controller movement, transmission of these digital signals to the patient's location, and electro-mechanical translation of these signals" (2005).

- **Lag.** The delay between movement in the operative field and the time a surgeon visually acknowledges such movement is referred to as a visual discrepancy, also commonly referred to as lag. Hanley et al., expands the meaning of lag by stating, "this time is the sum of the delays inherent to the digitalization and compression of the video signal(s) by the coder-decoder(s), transmission of the signal(s) across telecommunication networks and decompression of the signal(s) by the remote coder-decoders" (2005).

- **Round-trip delay.** The delay of both control latency and visual discrepancy, i.e., the time incurred between when the remote surgeon's hand movements at the telemanipulators and when such actions are visually acknowledged at the remote location (Hanley et al., 2005).

- **Bandwidth.** In computer networking, bandwidth is a term used to describe the data transmission rate designating the "maximum quantity of information that can be transmitted through a given communications circuit per unit of time" (Hanley et al., 2005). Satava states that the "addition of haptics also greatly increases the bandwidth needed for surgery, since there must be a sampling rate of nearly 1 kHz (1000 time a second) in order to provide a high-fidelity (consistent) response for the surgeon" (Satava, 2005). Haptics is the sense of human touch, and when applying the term to technology, haptics is explained by giving the user force feedback at the controls. An example of haptics as applied to the telemanipulators of a surgeon's console would be if a robotic arm were to rest on the chest of a patient, the telemanipulators would push back every time the patient's chest expanded while inhaling.

- **Jitter.** Jitter is the variation in the time between audio/video packets arriving, caused by network congestion, timing drift, and route changes. Hanley et al., states that "quality of service (QoS) indicates the degree of guarantee of or commitment to a particular quality of network service, including a defined minimum rate of data delivery (bandwidth), as well as the maximum intervals between information packet delivery" (2005). Notably sensitive to network communication delays, telesurgery is supremely influenced by this aspect. As one might expect, QoS is lowest over the public Internet where there is significantly more traffic than private/dedicated networks. QoS is at its highest when using dedicated communication assets such as Integrated Services Digital Network (ISDN) and Asynchronous Transfer Mode (ATM) networks (Hanly et al., 2005).

Technical and Legal Issues with Telesurgery

Telecommunications are not perfect. At one time or another, almost everyone has experienced a dropped call or had an internet connection go down, resulting in nothing more than a mere inconvenience or aggravation. However, in telesurgery, if a connection is lost, it could result in the patient being injured or worse. The question is, "Should the telecommunications company bear the liability, or should the liability ultimately be borne by the surgeon?" Issues such as this which could easily lead to legal problems as well as other problems associated with connectivity can lend themselves to the creation of a possible stigma associated with remote telesurgery.

Malpractice and Connectivity

The first glaring issue with telesurgery is the lack of face-to-face contact between doctor and patient which could ultimately become a determining factor in a malpractice suit. Whether creating a site-to-site Virtual Private Network (VPN) i.e., one's own direct link with no other traffic, or leasing a point-to-point ATM line for the use of telesurgery, it may involve more than one state or country. As a result of the transmission crossing any number of states and countries, conflicts regarding jurisdiction in a malpractice suit may arise. In addition, Marescaux et al., states that "other legal issues also need to be addressed, such as whether the surgeon should or not be liable for errors related to delays in transmission or equipment failure or whether a special consent should be obtained, and who is the person responsible for it" (2002). As previously stated, computer-assisted robotic telesurgery has the potential to provide expert healthcare to remote areas and developing countries that would not normally get such expertise since over 200 countries worldwide have access to the high-speed terrestrial backbone of the internet. Despite this fact, there continues to be portions of the globe that are not accessible through a terrestrial network and satellite communications, and these areas still face latency times around 1,000-14000ms round trip, which is nowhere near acceptable times allowed for telesurgery (Marescaux et al., 2002; Marescaux et al., 2001).

Dealing with Technical Issues

It is explained by Satava that "in the case of remote surgery, the limiting is not so much distance as it is latency, or the lag time between when the surgeon moves the handles of the console and the tip of the instrument moves in the remote location" (2005). These latency limitations are as follows. The surgeon perceives no latency if there is up to a 25ms delay in the connection (Satava, 2005). In the case that there is between a 25ms to 50ms delay the surgeon is aware that something is not correct but is able to compensate (Satava, 2005). "At 100ms, the effect of lag is very apparent, yet the surgeon is able to compensate through a number of strategies, such as moving slower and more deliberately or using a move-and-wait stepwise approach to the procedure" (Satava, 2005). Additionally, Satava states "when the delay is more than 200ms, it is extremely difficult (and in some systems impossible) to compensate, especially if force feedback is incorporated, which causes a conflict between forward intention and feedback that results in an unstable vacillation of the system" (2005).

First Telesurgery Procedure

September 7th, 1991, Professor Jacquese Marescaux, M.D., of the European Institute of Telesurgery (EITS) sat at a surgeon's console in a France Telecom/Equant center in New York City and performed the first transoceanic telesurgery on a sixty-eight year-old female in Strasbourg, France, over 4,000km away (Business Wire HealthWire, 2001; Leng, 2008; Marescaux, Leroy, Gagner, Rubino, Mutter, Fix, Butner, Smith, 2001). The robotic system Zeus, designed and manufactured by Computer Motion and later discontinued in 2003 after merging with Intuitive Surgical, was used to perform a laparoscopic cholecystectomy (gallbladder removal). The telesurgery was carried out without incident only "after obtaining ethical committee approval and informed consent from the patient" (Marescaux et al., 2001). The procedure took forty-five minutes to complete, during which time, forty technicians and medical staff were present in Strasbourg, France, including surgeons who were there to monitor the patient (Marescaux, 2001). New York City and Strasbourg, France, were connected via transatlantic optical-fiber network, which transported data at a bandwidth of ten megabits per second, through dedicated (private) connections using asynchronous transfer mode technology designed for real-time, low latency voice and video communication (Marescaux et al., 2001).

France Telecom took extra precautions as described by Tom Wyrick when he said, "It's always possible that an ocean cable can break or that some trawler (fishing vessel) will snag the cable. But that's why we installed a complete backup system that we could switch to within 10 seconds" (Larkin, 2001). Marescaux estimated that safe, acceptable 'round-trip delay' to be 330ms. With the mean delay time of 150ms, Marescaux was able to beat the estimated time by over 150ms, despite the fact that the video data traveled over 14,000km round-trip (Marescaux et al., 2001; Larkin, 2001). Operating and maintenance (OAM) packets were injected into the data stream, and were extracted in real time and analyzed by the network termination unit (NTU), ie., the receiver, and found that no packets were lost during transmission (Marescaux, J., Leroy, J., Rubino, F., Smith, M., Vix, M., Simone, M., Mutter, D., 2002). Following the completion of the surgery, three surgeons in New York were asked to give a subjective evaluation, with emphasis on image quality and the overall perception of safety of the procedure. Evaluations were on a zero-to-ten scale, with zero representing the worst possible and ten representing the best possible outcome. Image quality garnered an average score of nine and a half, and all three surgeons rated the perception of the safety of the operation, which Marescaux states "reflects the confidence of the surgeons and the reliability of the total system" as a prefect ten (2002).

Robot-Assisted Cholecystectomy

"Laparoscopic Cholecystectomy was one of the very first procedures to prove the applicability of surgical robots for general surgery" (Bodner, Hoeller, Wykypiel, Klinger, Schmid, 2005). The knowledge of this statement is of particular significance when one considers the following information.

Surgery.com states about twenty million Americans are living with cholelithasis, i.e., inflammation of the gallbladder (2012). Fortunately only two to three percent or about five hundred to six hundred thousand of the estimated twenty million need surgical intervention by means of cholecystectomy, i.e., removal of the gallbladder (Surgery.com, 2012). Initially, cholecystectomy cases were performed by laparotomy (open procedure), with the abdominal cavity exposed with a large incision. With the introduction of laparotics into the surgical treatment of cholecystectomies, complication rates have declined (as cited in Bass, Pitt, Lillemore, 1993; as cited in Begos, Modlin, 1994; Ruurda, Simmermacher, Borel Rinkes, Broeders, 2002). During the infancy years of computer-assisted robotic surgery Ruurda et al., conducted an evaluation of forty computer-assisted robotic laparoscopic cholecystectomies (2002). While it has since been concluded that computer-assisted robotic laparoscopic cholecystectomies are not justifiable for routine procedures, Ruurda et al. (2002), the study assessed "the feasibility of operating with a robotic telemanipulation system under well-controlled circumstances" (Jayaraman, Davis, Schlachta, 2008).

Patients and Methods

For the purposes of this evaluation, forty patients (F: 27, M: 13) were treated for cholelithasis through surgical intervention with the *daVinci* Surgical System between June 2000 and May 2001(Ruurda et. al., 2002). Gallstones were confirmed through ultrasound on all forty patients (Ruurda et. al., 2002). Gallstones are stone like formations that accumulate in the gallbladder and "are formed when the bile that is produced by the liver becomes very enriched with fatty substances" (USC.edu, 2012). Median age of the patients was forty-five years old, average weight was a hundred and seventy eight point two pounds and a body mass index of twenty-eight (Ruurda et. al., 2002). Thirty-five of the procedures were performed in a standard operating room with two of the three surgeons being experienced laparoscopic surgeons, along with standard operating room staff (Ruurda et. al., 2002). "One of the surgeons controlled the master console while one of the other surgeons assisted at the operating table in every case" (Ruurda et. al., 2002). The remaining five procedures were all performed in one day in a "day care operating room with a robot setup" "by a single surgeon (I.B), assisted by a specialized scrub nurse and a resident with extensive experience in robot-assisted surgery (J.R.)" (Ruurda et. al., 2002).

The procedures began with establishment of a pneumoperitoneum, i.e., filling the abdominal cavity with carbon dioxide (CO_2) gas, allowing for better visibility and maneuverability. Once the pneumoperitoneum was established, a "camera trocar" i.e. a trocar of either 5mm or 10mm in diameter with a camera attached was inserted through the sub-umbilical (below the belly button) position as referenced in the diagram to the right at position "1" (Ruurda et. al., 2002). Next, an incision was made in the left subcostal (below the rib cage) region as referenced in the diagram to the right at position "2" where the right robotic arm was positioned (Ruurda et. al., 2002). Following the positioning of the right

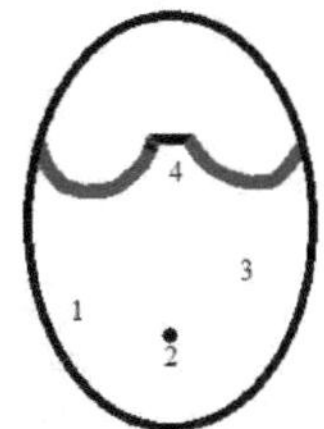

robotic arm, an incision was made in the right upper inguinal (above the groin) region as referenced in the diagram to the right at position "3" and the left robotic arm was positioned (Ruurda et. al., 2002). A fourth and final incision was made in the subxyphoid process (below the chest sternum) region as referenced in the diagram to the right at position "4" where a trocar was placed and a retractor would be introduced to retract the gallbladder (Ruurda et. al., 2002). The table-side surgeon was tasked with retracting the gallbladder while the surgeon at the console performed the cholecystectomy (Ruurda et. al., 2002).

Results

Of the forty computer-assisted robotic laparoscopic cholecystectomies, 39/40 or 98% were completed laparoscopically (Ruurda et. al., 2002). The single conversion to an open laparotomy was not due to robot deficiencies, but rather the surgeon's inability to retract the gallbladder due to severe cholelithasis (Ruurda et. al., 2002). As reported by Ruurda et al., "there were no robot-related surgical complications" but, "robot-related technical problems occurred in three cases" (Ruurda et. al., 2002). In all three incidents the "electrocautory instrument detached during the procedure" (Ruurda et. al., 2002). In two of the three incidents, the hook was able to be retrieved and removed laparoscopically, and in the third case a 4cm mini-laparotomy (open procedure) was performed due to the patient being "very obese" (Ruurda et. al., 2002).

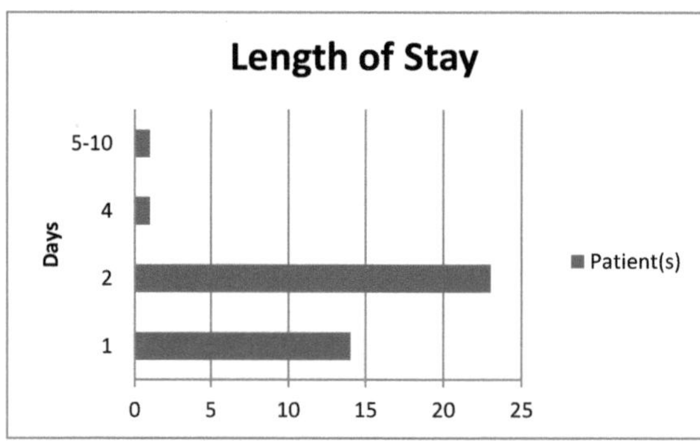

Source: Ruurda, P.J., Simmermacher, R.P.M., Borel Rinkes, I.H.M., Broeders, I.A.M.J., (2002). Robot Surgery in a Routine Procedure: *An Evaluation of 40 Robot-Assisted Laparoscopic Cholecystectomies.*

With the experience gained from previous cases, the median time was 15 minutes or less in the last 25 cases (Ruurda et. al., 2002). Median time from skin incision to closure of incisions was 82 minutes, with a range of 40 to 180 minutes (Ruurda et. al., 2002). With a range of 1 to 10 days, the total median time for hospitalization was 2 days (Ruurda et. al., 2002). Referring to the figure on the left, of the forty patients, 14 (35%) were dismissed postoperative day 1 and 23 (54%) were dismissed postoperative day 2(Ruurda et. al., 2002). The patient requiring the mini-laparotomy was hospitalized for 4 days and the patient who was converted to full open laparotomy was required to stay in the hospital 10 postoperative days (Ruurda et. al., 2002).

Discussion

"Although a relatively simple procedure like laparoscopic cholecystectomy is not a procedure in which surgeons benefit most from the advantages offered by a robotic telemanipulation system, it offered us the opportunity to assess the feasibility of working with this novel technology in a well-known and safe environment" (Ruurda et. al., 2002). The majority of patients (37\40 or 93%) were released by postoperative day 2, which is concurrent with empirical data (as cited in Cappuccino, Cargill, Nguyen, 1994; as cited in Scott, Zucker, Bailey, 1992; Ruurda et. al., 2002). During cases two, five, and six, robot-related technical problems occurred, and once the difficulties were identified and fixed, no further robot-related technical problems occurred (Ruurda et. al., 2002).

Long-Term Follow-up After Robotic Cholecystectomy

The long term results of computer-assisted robotic laparoscopic cholecystectomies (removal of the gall bladder) needs to be reviewed and considered as a measurement in determining the success of such surgery. The following case study "presents the first long-term results of robotic cholecystectomies" (Bodner et. al., 2005). Occurrences such as bile duct gallstones and bile duct strictures were investigated (Bodner et. al., 2005). The bile duct delivers digestive enzymes through a tube that runs from the gallbladder to the stomach. A bile duct stricture is a narrowing of the bile duct, preventing bile from draining into the intestine, and can cause liver failure if not treated (USC.com, 2012). "Because recurrent symptoms are the primary indication for cholecystectomy, quality of life after the operation was the major target of interest in this study (Bodner et. al., 2005).

Patients and Methods

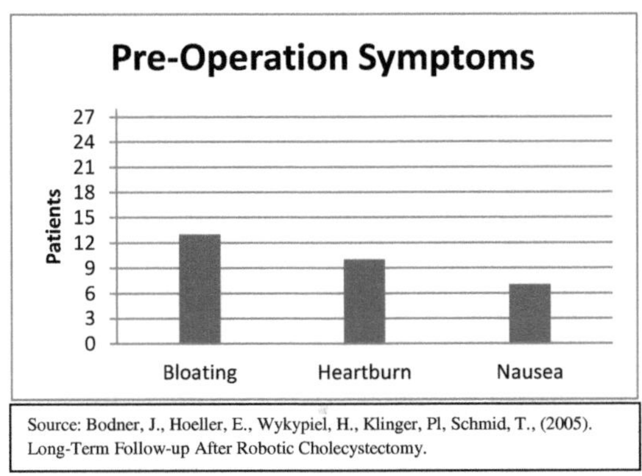

Source: Bodner, J., Hoeller, E., Wykypiel, H., Klinger, Pl, Schmid, T., (2005). Long-Term Follow-up After Robotic Cholecystectomy.

The first 25 robotic cholecystectomies with the *daVinci* surgical system were performed between June and November 2001 at Innsbruck University Hospital in Tyrol Austria (as cited in Bodner, Schmid, Wykypiel, Bodner, 2002; Bodner et. al., 2005). The method of selecting candidates was nonrandomized and depended heavily on the "availability of the robotic system and a surgeon trained in robotic surgery" (Bodner et. al., 2005). Informed consent was obtained from all patients and use of the *daVinci* Surgical System was approved by the ethics committee at Innsbruck University Hospital (Bodner et. al., 2005). Unlike the Ruurda et. al. study, indication for cholecystectomy was "symptomatic gallstone disease in all cases" (Bodner et. al., 2005). Due to robotic system failure, two of the cases had to be converted to a standard laparotomy, and as a result, the two cases were left out of the study because the criteria was compromised (Bodner, 2005). Bodner et. al., states that "patients were assessed on the basis of standardized follow up management including a clinical examination, blood tests, abdominal sonogram, and a quality-of-life questionnaire" (2005). The examination consisted of an assessment of scars, "as well as on the aspect and palpation of the right upper abdomen" (Bodner et. al., 2005). Palpation is the process of examination through the use of touch, or with a diagnostic aid. The abdominal ultrasound focused on the liver, pancreas and biliary tract (Bodner et. al., 2005). To evaluate each patient's subjective condition a "standardized symptoms questionnaire was modified for cholecystectomy after Bates" (as cited in Bates, Ebbs, Harrison, A'Hern, 1991; as cited in Wilson, Macintyre, 1993; Bates et. al., 2005). Along with other subjective questions regarding experience and frequency of listed symptoms, patients were asked to "assess the success of the operation in terms of "crude," "improved," "same," or "worse"" (Bodner et. al., 2005).

Results

Clinically controlled visits to patients were conducted 30 to 35 months after their respective procedures (Bodner et. al., 2005). Gallstone disease reoccurrence was found in one patient (4%), who suffered from choledocholithiasis i.e., presence of at least one gallstone in the bile duct, 29 months post robotic cholecystectomy (Bodner et. al., 2005). Following a clinical examination of the abdomen and inspection of scars, all 23 patients were deemed free of pathological findings (Bodner et. al., 2005). When asked about the appearance of their scars, all patients conveyed that they were pleased with the cosmetic result (Bodner et. al., 2005). The sonogram revealed the absence of "residual concrements" i.e., fragments of gallstones, in the abdomen (Bodner et. al., 2005). Of the 23 patients, 22 (96%) reported that their symptoms were either cured or significantly improved following the procedure (Bodner et. al., 2005). Bodner et. al., reported "the robotic approach was positively viewed by all 23 patients: 22 patients (96%) reported that they would opt for a robot-assisted procedure again if offered" (2005).

Discussion

"The introduction of laparoscopic cholecystectomy some 17 years ago heralded a new era in the treatment of gallbladder disease" and "since then, laparoscopic cholecystectomy has become the gold standard for symptomatic cholecystolithiasis" i.e., presence of gallstones in the gallbladder (Bodner et. al., 2005). In the opinion of Bodner et. al., limited vision and control are experienced during "complicated gallstone disease" cases (2005). By using computers, range of motion and dexterity are enhanced when compared to standard laparoscopic surgery (Bodner et. al., 2005). "No advantage of the robotic approach was found when compared to the conventional laparoscopic procedure" (as cited in Marescaux, Smith, Folscher et. al., 2001; as cited in Ruurda, Visser, Broeders, 2003; as cited in Nio, Bemelman, Busch, et. al., 2004; Bodner et. al., 2005). It was determined that there was no advantage found between the two aforementioned approaches, "no bile duct alteration were detected, and laboratory findings were unremarkable" (Bodner et. al., 2005). Reoccurrence of gallstone disease was found in one patient 29 weeks post operatively, but it was determined that there was no correlation between it and the surgical procedure (Bodner et. al., 2005). "Symptomatic outcome was also excellent" (Bodner et. al., 2005) This leads to the conclusion that patients undergoing robotic cholecystectomy can expect an excellent long-term prognosis (Bodner et. al., 2005).

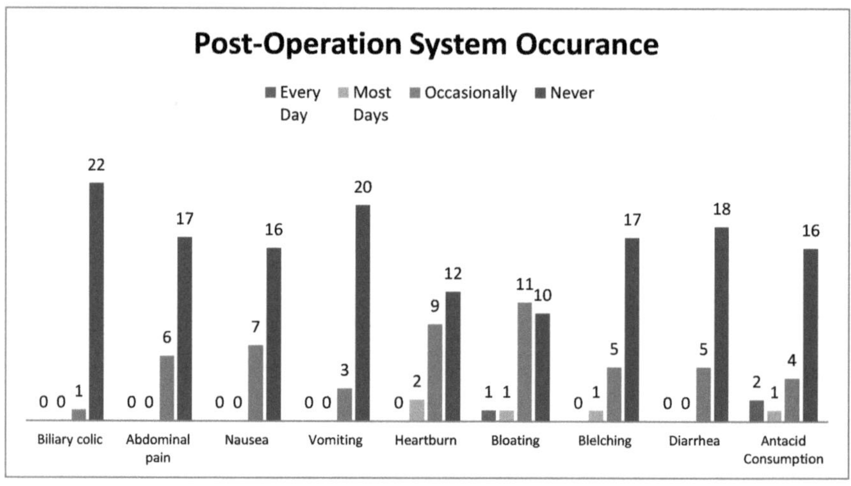

Telesurgery as an Educational Tool

The field of surgery and technology are one in the same in that they are continuously evolving. New technology is constantly being developed to outperform previous technology. Much like technology, the health care industry, too, is forever changing with experimental drugs, new procedures and techniques. As a result, the need to educate becomes increasingly important for those needing to remain abreast with industry standards and trends. Telesurgery offers a means to help teach and train students in surgical procedure, and as such, it can be used as a valuable educational tool.

Educating the Student

Patients are not the only ones to benefit from this new technology that is revolutionizing minimally invasive surgery as people know it. The potential is there to educate students by improving teaching and mentoring practices while reducing the learning curve associated with learning new procedures. The *daVinci* Surgical System can link two surgeon's consoles together with a special-purpose hardware/software connection, offering a microphone and a speaker that allows two-way communications during the surgical procedure if the consoles are not located remotely close to one another. (Hanly, E. J., Miller, B. E., Kumar, R., Hasser, C., Coste-Maniere, E., Talamini, M. A., Aurora, A. A., Schenkman, N. S., Marohn, M. R., 2006). When linked, an expert in the practice of surgery could complete the procedure by himself/herself while simultaneously remaining in constant communication with the student for teaching and mentoring purposes by helping with the identification of anatomical landmarks and structures. Numbers differ depending on which study one looks at, but a study conducted by the Hearst Corporation in 2009 "estimated 200,000 Americans will die needlessly from preventable medical mistakes and hospital infections" (2009). It is believed that with direct intervention of an expert, "remote telesurgery will likely reduce the errors that are caused by lack of experience or related to the early phase of the learning curve of new procedures" (Marescaux et al., 2002). Through these new teaching practices and training, it is believed that "telesurgery might eventually improve the standard of surgical care throughout the world" (Marescaux et al., 2002).

Recent Advances and Future Applications of Robotics in Health Care

One can only speculate as to what the future might hold in the way of technological advancements in the field of health care. The practice of remote telesurgery is relatively new, and as a result, much within the current scope of applications is merely theoretical. However, while many of the potential applications are theoretical, they are, nonetheless, within the realm of possibility, given the current rate of steady advancement in telecommunications. Furthermore, with the capabilities to perform a process remotely, the possibilities of more applications in remote health care, in general, are endless in number. The following provides a brief look at a couple of recent advances using robotics in medicine as well as a glimpse into possible future applications of computer-assisted robots in telesurgery.

The Penelope™

Using robots to operate on humans is something that one once would have found only in a science fiction book or movie, and now, that technology is being used daily. It is hard to imagine what the future of health care will be for mankind, but some of what the future potentially holds has already started making its way into the health care setting. "The Penelope™ Surgical Instrument Server is the world's first autonomous, vision-guided robotic surgical assistant" (Robotic Systems & Technologies, 2010). Penelope is outfitted with a "sophisticated object recognition camera system that can identify various surgical instruments" (Satava, 2005). Using voice recognition, Penelope listens for the surgeon's command for an instrument, and using its robotic arm, which offers five degrees of freedom along with a stationary camera, locates the instrument, retrieves it and extends it to the surgeon (Robotic Systems & Technologies, 2010; Satava, 2005). The surgeon then drops the instrument in his/her hand and retrieves the one from Penelope, "which then searches for a dropped instrument and returns it to the scrub table – all in a fashion identical to current scrub nurse practice" (Satava, 2005). As stated by Satava, "clinical trials have shown over 96% accuracy in instrument identification with no unexpected delays or errors" (2005).

Robotic Rounds

Yulun Wang has developed a robot that enhances another aspect of health care that is common to medical physicians, and that is making rounds on patients, whether it is preoperative, post operative, morning, nursing or other (Satava, 2005). Louis Kavoussi, a urologist at John Hopkins, uses the robot Wang developed to perform remote "postoperative rounds from his office to the urology ward in the hospital" (Satava, 2005). "This system is a teleoperated mobile robot (actually a standard televideo system mounted on a mobile robot platform with a flat-panel display for a "head") which the physician can control from their office, clinic, or other place where a televideo system with joystick controller is located" (Satava, 2005). According to Satava, initial implementation has yielded surprising reviews and a "very high satisfaction rate from patients" (2005). Satava goes on to state that "it was discovered that patients like talking to Kavoussi on the mobile robot because they have his undivided attention, something that is not always possible during rushed postoperative rounds with interruptions from an intern, resident, or nurse" (2005). It has been reported by the patients that the 'eye-to-eye' (through the use of a video monitor) contact provides a sense of "focus and attention and communication from the physician" (Satava, 2005). Although this can supplement bedside care, patients still do not get the same safe feelings as they do from a direct doctor-to-patient relationship during face-to-face rounds. It can be said then that in telemedicine, certain aspects can be replicated in providing health care, but some of the basic needs of a human can only be satisfied with direct contact.

Future Applications of Computer-Assisted Robots in Telesurgery

The potential benefits from the application of remote robot-assisted telesurgery are multiple in numbers, but they are currently limited to only certain regions of the world. The terrestrial fiber optic backbone to the Internet has progressed rapidly since the introduction of computer-assisted robotic telesurgery due to the constant investment into telecommunications by municipalities and private carriers throughout the world. With the fiber optic backbone to support it, this relatively new technology is employed in nearly 2,000 hospitals worldwide, and as stated by Marescaux et al., "geographic constraints will no longer determine the type of treatment the patient receives because of lack of surgical expertise" (2002). In developing third world countries where health care is routinely provided by volunteers who may not necessarily be the most proficient in all fields of medicine and surgery, this technology could have an immeasurable impact. In addition, Marescaux et al., states "emergency operations in small rural hospitals are sometimes challenging for young surgeons on call" (2002). Terrestrial networks that connect small rural hospitals to larger metropolises where expert surgeons are more plentiful would allow them to assist or carry out the procedure themselves. This technology could also be employed on the battlefield, allowing military surgeons to be located away from the front lines while still providing their services to fellow service men and women. As technical obstacles are overcome, telesurgery could extend its services to oceanic submarine vessels and even into outer space, where astronauts could be operated on by "earthbound interventionalists" (Hanly et al., 2005).

Conclusion

Advances in technology in the 20th and 21st centuries have grown at an extraordinary rate, and these technological advances have both revolutionized and forever changed the manner in which people conduct their daily lives. Among these advances, those in the area of medicine have probably had the most profound and far-reaching effect for people in terms of improving their quality of life. Throughout their lives, most people have something happen to themselves, a family member, or a friend which requires medical attention and ultimately the need for surgical intervention. In the fast-paced society in which people live today, people want to return to living their daily lives as quickly as possible, and the use of computer-assisted robotic surgery offers them the least invasive surgical approach with the best possible outcome should surgery be required. As technology and telecommunications progress, people will see the use of telemedicine and telesurgery become even more entrenched in health care practice. Who knows phenomenal breakthroughs and medical advancements are just beyond the horizon? Only time alone will reveal what remarkable innovations mankind will see in the future which include computer-assisted robotics and telesurgery in the practice of medicine.

<u>Sources</u>

American Telemedicine Association, (2012). About Telemedicine. Retrieved from http://www.americantelemed.org/i4a/pages/index.cfm?pageID=3308

Bass, E.B., Pitt, H.A., Lillemore, K.D., (1993). Cost-effectiveness of Laparoscopic Cholecystectomy Versus Open Cholecystectomy. *Am J Surg. 165 (1993). 466-471.*

Bates, T., Ebbs, S.R., Harrison, M., A'Hern, R.P., (1991). Influence of Cholecystectomy on Symptoms. *Br j Surg. 78 (1991). 964-7*

Begos, D.G., Modlin, I.M., (1994). Laparoscopic Cholecystectomy: *From Gimmick to Gold Standard. J Clin Gastroenterol. 19 (1994). 325-330.*

Bodner, J., Hoeller, E., Wykypiel, H., Klinger, Pl, Schmid, T., (2005). Long-Term Follow-up After Robotic Cholecystectomy. *The American Surgeon. 71 (01/05). 281-285*

Bodner, J., Schmid, T., Wykypiel, h., Bodner, E., (2002). First Experience with the daVinci System. *Chirurg. 34 (2002).166-9*

Business Wire HealthWire, (2001). World's First Complete Telesurgery Procedure Performed Using Surgical Robots and Telecommunications Solutions With High-Speed Services. Retrieved from http://www.thefreelibrary.com.

Cappuccino, H., Cargill, S., Nguyen, T., (1994). Laparoscopic Cholecystectomy: *563 cases at a Community Teaching Hospital and a Review of 12,201 Cases in the Literature. Monmouth Medical Center Laparoscopic Cholecystectomy Group. Surg Laparosc Endosc. 4 (1994) 213-221*

Computer Motion, Inc. (2002). Retrieved from http://www.computermotion.com.

Coxworth, B., (2010). Surgical Robot Provides Haptic Feedback to Users. Retrieved from http://www.gixmag.com/sofie-force-feedback -surgical-robot/16613/.

Gadaez, T.R., (1993). Update on Laparoscopic Cholecystectomy, Including a Clinical Pathway. *Am J Surg. 165(1993) 450-454*

Gerhardus, D., (2003). Robot-Assisted Surgery: *The Future is Here. Journal of Healthcare Management 48(07/03), 242-251.*

Hanly, E. J., Marohn, M. R., Schenkman, N. S., Miller, G. R., Marchessault, R., Broderick, T. J., (2005). Dynamics and Organizations of Telesurgery. *European Surgery 37(08/05). 274-278.* doi: 10.1007/s10353-005-0181-0

Hanly, E. J., Miller, B. E., Kumar, R., Hasser, C., Coste-Maniere, E., Talamini, M. A., Aurora, A. A., Schenkman, N. S., Marohn, M. R., (2006). Mentoring Console Improves Collaboration and teaching in Surgical Robotics. *Journal of Laparoendoscopic & Advanced Surgical Techniques. 16(10/06), 445-451*

Hawaleshka, D. (2002). TECHNICAL MARVELS. (Cover story). *Maclean's, 115*(12), 48.

Hearst Communications Inc. (2009). Hearst National Investigations Finds Americans are Continuing to Die in Staggering Numbers From Preventable Medical Injuries. Retrieved from http://www. hearst.com/press-room/pr-20090809b.php

Heikkinen, T.J., Haukipuro, K., Bringman, S., (2000). Comparison of Laparoscopic and open Nissen Fundoplication 2 Years after operation. *A prospective randomized trial. Surg Endosc. 14(2000) 1019-1023*

Hirano, Y., Ishikawa, N., Watanabe, G., (2010). Suture Damage after Grasping with EndoWrist of the *daVinci* Surgical System. *Minimally Invasive Therapy, 19(2010), 203-206.* doi: 10.3109/13645706.2010.496951

Intuitive Surgical, Inc., (2012). EndoWrist Instruments. Retrieved from http://www.intuitivesurgical.com/products/

Jacobsen, G., Berger, R., Horgan, S., (2003). The role of Robotic Surgery in Morbid Obesity. *Journal of Laparoendoscopic & Advanced Surgical Techniques, 13(10/03), 279-283.*

Jayaraman, S., Davies, W., Schlachta, C.M., (2009). Getting Started with Robotics in General Surgery with Cholecystectomy: *The Canadian Experience. Can J Surg. 52 (05/09). 374-378*

Kaynan, M. A., (2009). Advantages of Robotic Surgery. Retrieved from http://www.roboticurologicsurgery.com/advantages-of-robotic-surgery.php.

Konsten, J., Gouma, D.J., von Meyenfeldt, M.F., Menheere P., (1993). Long Term Follow-up After Open Cholecystectomy. *Br J Surg. 80 (1993).100-2*

Larkin, M., (2001). Transatlantic, Robot-Assisted Telesurgery Deemed a Success. *The Lancet 358(09/01). 1074*

Lee, A. C. W., Harada, N., (2011). Telehealth as a Means of Health Care Delivery for Physical Therapist Practice. *Phys Ther. 92 (2012). 463-468*

Lee, S., Broderick, T. J., Haynes, J., Bagwell, C., Doarn, C. R., Merrill, R.C., (2003). The Role of low-bandwidth Telemedicine in Surgical Prescreening. *Journal of Pediatric Surg. 38(2003). 1281-1283.*

Leng, L., (2008). Surgery with Robots. *InnovationMagazine.com. 8(11/08). 65-67*

Marescaux, J., Leroy, J., Gagner, M., Rubino, F., Mutter, D., Vix, M., Butner, S. E., Smith, M. K., (2001). Transatlantic Robot-assisted Telesurgery: ATM Technology now Enables Operations to be Performed Over Huge Distances. *Nature 413(09/01). 379-380.*

Marescaux, J., Leroy, J., Rubino, F., Smith, M., Vix, M., Simone, M., Mutter, D., (2002). Transcontinental Robotic-Assisted Remote Telesurgery: Feasibility and Potential Applications. *Ann Surg. 235(04/02). 487-492.*

Meadows, M., (2002). Robots Lend a Helping Hand to Surgeons. *FDA Consumer, 36(01/02). 10-15.*

Murphy, G. D., Hall, R., Tong, R., Goel, R., Costello, J. A. (2008). Robotic Technology in Surgery: Current Status in 2008. *ANZ Journal of Surgery, 78(02/08), 1076-1081*. doi: 10.1111/j.1445-2197.2008.04754.x

Newlin, M. E., Mikami, D. J., Melvin, W. S., (2004). Initial Experience with the Four-Arm Computer-Enhanced Telesurgery Device in Foregut Surgery. *Journal of Laparoendoscopic & Advanced Surgical Techniques. 14(10/04). 121-124.*

Nio, D., Bemelman, W.A., Busch, O.R., et al. (2004). Robot-assisted Laparoscopic Cholecystectomy Versus Conventional Laparoscopic Cholecystectomy: *A Comparative Study. Surg Endosc. 18. (2004). 379-82*

Robotic Systems & Technologies, Inc., (2010). Penelope: 'Greetings, Doctor'. Retrieved from http://www.roboticsystech.com/productspenelope.html

Rodas, E. B., Latifi, R., Cone, S., Broderick, T. J., Doarn, C. R., Merrill, R. C., (2002). Telesurgical Presence and Consultation for Open Surgery. *Arch Surg 137(2002). 1360-1363.*

Ruurda, J.P., Visser, P.L., Broeders, I.A., (2003). Analysis of Procedure Time in Robot-assisted Surgery: *Comparative Study in Laparoscopic Cholecystectomy. Comput Aided Surg. 8 (2003). 24-9*

Ruurda, P.J., Simmermacher, R.P.M., Borel Rinkes, I.H.M., Broeders, I.A.M.J., (2002). Robot Surgery in a Routine Procedure: *An Evaluation of 40 Robot-Assisted Laparoscopic Cholecystectomies. Eur. Surg. 34 (2002). 170-172*

Satava, R. M., (2005). Telesurgery, Robotics, and the Future of Telemedicine. *European Surgery 37(08/05). 304-307.* doi: 10.1007/s10353-005-0185-9

Scott, T.R., Zucker, K.A., Bailey, R.W., (1992). Lararoscopic Cholecystectomy: *A Review of 12,397 Patients. Surg Laparosc Endosc. 2. (1992) 191-198*

Sharp, P. A., (2001). Surgical History: The History of Bloodletting. *ANZ J. Surg. 71(2001). A91-A92.*

Smith, A. C., Bensink, M., Armfield, N., Stillman, N., Caffery, L., (2005). Telemedicine and Rural Health Care Applications. *Journal of Postgraduate Medicine. 51 (2005). 286-293*

Stark, K., (2002). Robots Help Pennsylvania Surgeons to Perform Minimally Invasive Procedures. *The Philadelphia Inquirer.* Retrieved from http://www.philly.com.

University of Southern California, Department of Surgery, (2012). Bile Duct Strictures. Retrieved from http://www.surgery.use.edu/divisions/tumor/pancreasdiseases/ web%20pages/BILIARY%   20SYSTEM/bile%20duct%20strictures.html

Ure, B.M., Troidl, H., Spangenberger, W., et. al., (1995). Long-term Results After laparoscopic Cholecystectomy. *Br J Surg. 82 (1995).267-70*

Value and Feasibility, (2002). Tele-Surgery. Retrieved from
http://www.stanford.edu/~mdpatter/MIA/value_and_feasibility.htm.

Virtual Solder Project, (2011). Holomer. Retrieved from
http://www.virtualsoldier.us/holomer.htm

Whitten, P., Mair, F., (2004). Telesurgery versus Telemedicine in Surgery – an Overview. *Surgical Technology Int. 12 (2004). 68-72.*

YOUR KNOWLEDGE HAS VALUE

- We will publish your bachelor's and master's thesis, essays and papers

- Your own eBook and book - sold worldwide in all relevant shops

- Earn money with each sale

Upload your text at www.GRIN.com and publish for free